Unidentified Submerged Objects and Underwater Bases

There are many books on UFOs and Aliens but most are in the air or over the land.

However, you might not be aware that there are also many UFO sightings underwater, or over water.

There are even some strong claims of UFO and/or Alien bases underwater.

Over 71% of the surface of the Earth is covered in water, much of which is in the Earth's Oceans. So it makes sense that Aliens visiting Earth would hide in the water or even build bases there.

Yes—there are reports from several places of possible Aliens bases beneath the waves. This would make a sensible place to hide if you have advanced technology which will allow you to hide underwater.

Some of the suspected underwater bases are near Malibu-California, Gulf Breeze-Florida, and the Solomon Islands in the South Pacific. There are probably a lot more submerged base locations we don't know about all over the world.

In this book I've provided as many Unidentified Submerged Object stories as I could research. I think you will find these stories to be very interesting.

Unidentified Submerged Objects and Underwater Bases

Unidentified Submerged Objects and Underwater Bases

Copyright Page

The book is copyrighted for 2021

Unidentified Submerged Objects and Underwater Bases

The Aliens and UFO Secrets Series-Book Three

By Martin K. Ettington

All Rights Reserved USA 2021

ISBN: 9798736192595

Printed in the United States of America

Unidentified Submerged Objects and Underwater Bases

Unidentified Submerged Objects and Underwater Bases

Other books by Martin K. Ettington

Spiritual and Metaphysics Books:
Prophecy: A History and How to Guide
God Like Powers and Abilities
Enlightenment for Newbies
Removing Illusions to Find True Happiness
Using the Scientific Method to Study the Paranormal
A Compendium of Metaphysics and How to Guides (Six books together in one volume)
Love from the Heart
The Enlightenment Experience
Learn Your Soul's Purpose
Pursuing Enlightenment
A Modern Man's Search for Truth
Use Intuition and Prophecy to Improve Your Life
The Handbook of Spiritual and Energy Healing

Longevity & Immortality:
Physical Immortality: A History and How to Guide
The Commentaries of Living Immortals
Records of Extremely Long Lived Persons
Enlightenment and Immortality
Longevity Improvements from Science
The 10 Principles of Personal Longevity
Telomeres & Longevity
The Diets and Lifestyles of the Worlds Oldest Peoples
The Longevity Six Books Bundle

Science Fiction:
The History of Science Fiction and Fantasy
Out of This Universe
Personal Freedom-Parts 1 & 2
The Psychic Soldier Series:
　Book 1-Himalayan Journey
　Book 2-A Soldier is Born
　Book 3-Fighting For Right
Book 4-Earth Protector
The Immortality Sci Fi Bundle

The God Like Powers Series:
Human Invisibility
Invulnerability and Shielding
Teleportation
Psychokinesis
Our Energy Body, Auras, and Thoughtforms
The God Like Powers Series—Volume 1 Compilation

The Yoga Discovery Series:
Yoga-An Ancient Art Form
Hatha Yoga-Helping you Live Better
Raja Yoga-Through the Ages
The Yoga Discovery Package

Business & Coaching Books:
Creating, Paublishing, & Marketing Practitioner Ebooks
Building a Successful Longevity Coaching Business
Why Become a Coach?
The Professional Coaching Success Trilogy
2020-Make Money Writing and Selling Books
The 2020 Handbook of High Paying Work Without a College Degree

Science, Technology, and Misc.
Future Predictions By and Engineer & Seer
The Unusual Science & Technology Bundle
The Real Atlantis-In the Eye of the Sahara
Removing Limits On Our Consciousness-And Thinking Outside the Box

Unidentified Submerged Objects and Underwater Bases

Legendary Animals and Creatures
Are Cryptozoological Animals Real or Imaginary?
Fire in History and Mythology
All About Dragons
Sea Serpents and Ocean Monsters
The Importance of Fire in History and Mythology

Ancient History
The Real Atlantis-In the Eye of the Sahara
Ancient & Prehistoric Civilizations
Ancient & Prehistoric Civilizations-Book Two
The History of Antediluvian Giants
The Antediluvian History of Earth
Ancient Underground Cities and Tunnels
Strange Objects Which Should Not Exist
More Out of Place Artifacts
Strange and Ancient Places in the USA
A Theory of Ancient Prehistory And Giant Aliens

Aliens and Space
Aliens and Secret Technology
Aliens Are Already Among Us
Unidentified Submerged Objects and Underwater Bases

Living in Space
Designing and Building Space Colonies
Humanity and the Universe
All About Moon Bases
All About Mars Journeys and Settlement
The Space and Aliens Six Books Bundle
A Theory of Ancient Prehistory and Giant Aliens
The Space Colonies and Space Structures Coloring Book
All About Asteroids
Spaceships, Past, Present, and Future
Astronauts, Cosmonauts, and Other Important Space Flyers
All About Mars Journeys and Settlement
Mining the Asteroid Belt
Exploring and Settling Our Huge Solar System

Survival
33 Incredible True Survival Stories
How to Survive Anything: From the Wilderness to Man Made Disasters
Building and Stocking a Nuclear Shelter for less than $10,000
Survival of Humanity Throughout the Ages

Time Travel
Real Time Travel Stories From a Psychic Engineer
The Real Nature of Time: An Analysis of Physics, Prophecy, and Time Travel Experiences

Unidentified Submerged Objects and Underwater Bases

<u>The Longevity Training Series</u>

(A transcription of the online Multimedia Longevity Coaching Training Program)
The Personal Longevity Training Series-Book1-Long Lived Persons
The Personal Longevity Training Series-Book2-Your Soul's Purpose
The Personal Longevity Training Series-Book3-Enable Your Life Urge
The Personal Longevity Training Series-Book4-Your Spiritual Connection
The Personal Longevity Training Series-Book5-Having Love in Your Heart
The Personal Longevity Training Series-Book6-Energy Body Health
The Personal Longevity Training Series-Book7-The Science of Longevity
The Personal Longevity Training Series-Book8-Physical Body Health
The Personal Longevity Training Series-Book9-Avoiding Accidents
The Personal Longevity Training Series-Book10-Implementing These Principles
The Personal Longevity Training Series-Books One Thru Ten

These books are all available in digital and printed formats from my website and on Amazon, Barnes & Noble, Apple ITunes, and many other sites. My Books Website is: http://mkettingtonbooks.com

Unidentified Submerged Objects and Underwater Bases

Signup for our Mailing List to get the following:

1) A discount coupon for 25% discount on all books on our site

2) Occasional Notices of new books available

3) Occasional Email on other offerings of ours (Monthly)

Go to this link to sign-up:

http://personal-longevity.com/mkebooks/emailsignup/

And click this link to get the FREE 102 page Ebook titled "Secrets of Many Things"

If you have any questions about this book or other subjects please contact the Author at:

mke@mkettingtonbooks.com

Unidentified Submerged Objects and Underwater Bases

Table of Contents

1.0 Introduction 1
2.0 Underwater Alien Bases...3
  2.1 Malibu, California, USA Underwater Base..................3
  2.2 Gulf Breeze Florida, USA 7
  2.3 Solomon Islands Underwater Bases 11
  2.4 Underwater Base Off Puffin Island Wales 13
  2.5 UFO Base Under Lake Ontario 15
  2.6 Flight Corridor To Underwater Base North Island, New Zealand..17
  2.7 Pacific Ocean 'Humming' Might be USOs...........19
  2.8 Strange Lights and Sounds In Puerto Rico 21
  2.9 Ancient Alien Base at Lake Titicaca 23
  2.10 Underwater Alien Base at Guantanamo Bay 25
  2.11 Increased UFO Sightings at Lake Erie Cleveland, Ohio..27
  2.12 Lake Baikal—Scene of an Underwater Battle with Aliens 29
3.0 Unidentified Submerged Objects 31
  3.1 The Utsuro Bune Story 31
  3.2 Christopher Columbus and the light sightings..........37
  3.3 Puerto Rico Sightings 41
  3.4 Japanese Fishing Boat 45
  3.5 Another Sighting off Puerto Rico................................47

Unidentified Submerged Objects and Underwater Bases

3.6 Half Moon Bay, California...49

3.7 Imperial Beach, California 51

3.8 Nuclear Submarine Stories 53

3.9 Lake Baikal, Russia..57

3.10 Russian Subs and USO Battle 69

3.11 UFO Crash into Water over Hawaii 73

3.12 USOs at Guantanamo Base, Cuba 75

3.13 Red Sphere in the Red Sea......................................79

3.14 The Baltic Sea Anomaly 81

3.15 Silver Beehive 83

3.16 The Bermuda Triangle...85

3.17 The Devil's Sea..87

3.18 Pyramid USOs over a Destroyer.............................89

4.0 Other Interesting Stories- The Swimmers...................93

5.0   Summary 95

6.0 Bibliography...97

Unidentified Submerged Objects and Underwater Bases

1.0 Introduction

There are many books on UFOs and Aliens but most are in the air or over the land.

However, you might not be aware that there are also many UFO sightings underwater, or over water.

There are even some strong claims of UFO and/or Alien bases underwater.

Over 71% of the surface of the Earth is covered in water, much of which is in the Earth's Oceans. So it makes sense that Aliens visiting Earth would hide in the water or even build bases there.

Yes—there are reports from several places of possible Aliens bases beneath the waves. This would make a sensible place to hide if you have advanced technology which will allow you to hide underwater.

Some of the suspected underwater bases are near Malibu-California, Gulf Breeze-Florida, and the Solomon Islands in the South Pacific. There are probably a lot more submerged base locations we don't know about all over the world.

In this book I've provided as many Unidentified Submerged Object stories as I could research. I think you will find these stories very interesting.

This book complements by other books "Aliens and Secret Technology: A Theory of the Hidden Truth" and "Aliens are Already Among Us"

Unidentified Submerged Objects and Underwater Bases

Unidentified Submerged Objects and Underwater Bases

2.0 Underwater Alien Bases

Underwater Alien Bases are very possible. Consider two things:

A) Aliens with the technology to reach Earth over interstellar distances should also be able to build ships which can fly and exist underwater too.

B) With 71% of the world covered with water using underwater locations to hide is an easy thing to do.

2.1 Malibu, California, USA Underwater Base

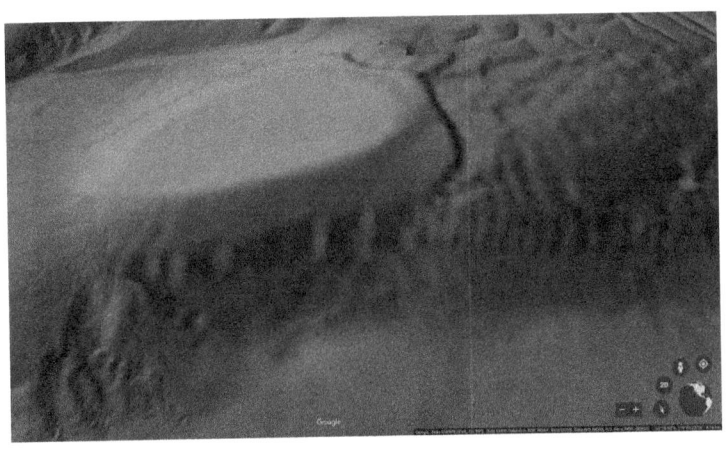

LOCATION: About six miles off the coast of Malibu's Point Dume, in California

For decades, reports of UFO (unidentified flying object) and USO (unidentified submerged object) sightings along the Malibu coast have been fairly common. Rumors of an alien base, or portal, in this area are nothing new, but in recent years, they have come to the fore within the UFO research community. The attention, specifically to this location, is largely due to reports of a stadium-shaped "structure" about 2,000 feet underwater, widely known as Sycamore Knoll. Some refer to it as an anomaly while others believe is an alien base. It's believed to be between two-and-a-half and three-miles wide.

In 2014, radio personality Jimmy "Captain of Conspiracy" Church brought the matter up on his show *Fade to Black*, referring to the area as a "hub of USO/UFO activity going back 40 years." Church questioned whether it could be natural or proof of an entrance to something inland—perhaps another Area 51 (the massive aircraft-testing facility in the Nevada desert that has long been the subject of UFO theories and investigations, even after more solid explanations were provided by the government)—or something not yet considered. He said it was "potentially the biggest breaking story in ufology since Area 51."

But not everyone in the UFO research community was convinced. Robert Stanley, a UFO researcher and the editor of *Unicus* magazine, the self-described "magazine for earthbound extraterrestrials," reports having noticed this mysterious spot via Google Earth several years prior to Church's announcement, but says he didn't report it publicly as he was unable to substantiate its authenticity. Stanley says he doesn't know what the anomaly is, but he does offer a few thoughts. "It feels like a red herring," he says. "This is some sort of weird distraction. It may be because Point Mugu [which is nearby] is a very large, powerful naval base that they don't want people to think of in terms of [it being a] secret military base."

Preston Dennett, another UFO researcher and author of more than 20 books on paranormal and supernatural phenomena, is less skeptical. He's believed in the alien base theory for years. "I was convinced there was something there before this Malibu anomaly was publicized," he says.

PARANORMAL ACTIVITY: Opinions are mixed about what the anomaly could be. Is it an alien base? Is it a government-created distraction? Or is it just the shape of the underwater terrain in the depths of the Pacific? Stanley stands firm on his nope-to-an-alien-base stance. "A lot of people are under the false impression that there's a UFO base there," he says. "Now that doesn't mean that there aren't UFOs there. I've seen them myself—on many occasions. But just because UFOs are seen there, that doesn't mean that that is a UFO base under the water."

And here's his reasoning: "[UFOs] are coming in and out of portals, star gates or wormholes. They don't need massive bases under the water or under the ground. I'm not saying there aren't any, but they don't need them because they can just pop in and out of our skies anytime they want, anywhere they want, as far as I can tell." Dennett approaches the topic from the other side, saying, "I've collected probably 50 or 100 reports of people who have seen UFOs going in and out of the water there." Still, he expresses some doubt: "My problem is that the Google images are coming out different, depending on what viewpoint you're looking at this thing from. Some show a tunnel. Some don't."

While he acknowledges he hasn't gone down to explore the structure, Dennett remains optimistic about it being something extraterrestrial. "I'm very intrigued by the possibility," he says. "I'm convinced there's something down there."

Unidentified Submerged Objects and Underwater Bases

Among the many discussions of this mysterious area over the years, Dennett's and Stanley's coverage remain at the forefront.

Dennett's 1999 book *UFOs Over Topanga Canyon* addresses accounts of UFO sightings from the Malibu area in general, and he presented the possibility of the underwater anomaly being a base as far back as 2006, when *Fate* magazine published his article **"Is There an Underwater UFO Base off the Southern California Coast?,"** where he cited several witness accounts and opinions.

In several articles for Unicus, including **"Who Discovered the So-Called Malibu UFO Base?"** .

SEE FOR YOURSELF: Take a look at **Google Earth** coordinates 34° 1'23.31"N 118° 59'45.64"W.

Unidentified Submerged Objects and Underwater Bases

2.2 Gulf Breeze Florida, USA

Alien spaceship may lay off Florida coast, says Discovery Channel treasure hunter

In a close encounter of the submerged kind, a Discovery Channel treasure hunter said he found what could be a massive extraterrestrial structure at the bottom of the Atlantic Ocean off the Florida coast.

Darrell Miklos, host of Discovery's treasure hunting show *Cooper's Treasure*, was chasing the trail of an English shipwreck for an episode of his show when he made the find 300 feet beneath the Bermuda Triangle. "I was trying to identify shipwreck material based on one of the anomaly readings on Gordon's charts when I noticed something that stuck out, that shocked me," Miklos told the *Daily Mail*. "It was a formation unlike anything I've ever seen related to shipwreck material, it was too big for that. It was also something that was completely different from anything that I've seen that was made by nature."

Unidentified Submerged Objects and Underwater Bases

For two seasons, Miklos has used maps made by his friend, the late NASA astronaut Gordon Cooper, to hunt for shipwrecked treasure around the Caribbean Sea. In the 1960s, Cooper flew missions for the Mercury and Gemini space programs. He died in 2004, but, Miklos said, not before passing to him an alleged treasure map. Miklos said Cooper was spying on potential Russian nuclear sites based on magnetic anomalies when he used the same anomalies to make his treasure map. Cooper was also a fervent believer in extraterrestrial life, claiming to have witnessed UFOs during his time in the Air Force.

Miklos' find, however, is a USO, an unidentified submerged object. The coral-covered structure features 15, 300-foot-long arms jutting out from a center object. Scientists on Miklos' team said the coral could be thousands of years old. They also said coral couldn't naturally grow in those formations.

"There's identical formations in three different areas and they don't look nature made, they don't look man made, certainly nothing I've ever seen based on my experience and I have years of experience at doing this," Miklos said. "We've identified multiple different types of shipwreck material, this doesn't match or look anything like that." Miklos and the Cooper's Treasure production company, AMPLE Entertainment, hope the find will urge the Discovery Channel to pick up the show for a third season so they can further explore the find.

"Cooper was a reliable source for treasure, then based on his findings Darrell found something that does not appear to be a shipwreck or anything that anybody has ever seen," AMPLE founder Ari Mark said. "We want to find out exactly what it is and establish whether it ties in with

Unidentified Submerged Objects and Underwater Bases

While surprised by the structure, Miklos said he want to investigate further before saying exactly what he thinks the structure could be.

He's not saying it was aliens? But … well, you decide.

Unidentified Submerged Objects and Underwater Bases

Unidentified Submerged Objects and Underwater Bases

2.3 Solomon Islands Underwater Bases

Solomon Islands Mysteries: Accounts of Giants and UFOs in the Solomon Islands

Near where the sunken warships of the Battle of Guadalcanal lie glowing UFOs rise out of the Pacific fly into the mountains and disappear into jungle lakes. Here a tropical paradise exists with inexplicable ancient ruins and puzzling writings of an unknown culture. Steamy rugged mountain ranges are inhabited by strange Sasquatch-like creatures. They have come down to the villages to kidnap the locals for generations. Terrifying stories of abduction and cannibalism are passed on by the villagers to their children. These are some of the incredible tales that the Solomon Islanders have lived with for decades and you will read about in this spellbinding book.

Author Marius Boirayon is the son of the World War II central France maquis (resistance) leader and grew up in Mount Hagen in the Papua New Guinea Highlands. Following a career in the Royal Australian Air Force and as

Unidentified Submerged Objects and Underwater Bases

an aircraft/helicopter engineer working in outback Australia he decided in 1995 to go to the Solomon Islands to live.

Unidentified Submerged Objects and Underwater Bases

2.4 Underwater Base Off Puffin Island Wales

Numerous sightings of strange lights leaving and entering the sea around Puffin Island, Wales, has led some ufologists to believe that the area may be home to an underwater alien base. A spate of sightings in early 1974 were particularly interesting to investigators. All spoke of not just lights, but solid objects that were seen leaving the sea near the island.

UFO investigator Phil Hoyle stated that he has heard and read various unconnected reports concerning Puffin Island and that all of them tell the same story and describe the same type of phenomenon. He also stated that alleged alien abductees close to the area have reported that their abductors were humanoid and told them they came from a base under the sea near the island.

There are even some theories that the area itself may be the ancient legendary kingdom of Cantre'r Gwaelod and that these humanoids are their descendants.

Unidentified Submerged Objects and Underwater Bases

Hoyle believes that the ties to an ancient site and the current UFO activity shouldn't be downplayed, stating that according to his research, there is an 80 percent increase in such activity on or around ancient sites worldwide.

2.5 UFO Base Under Lake Ontario

In December 2013, MUFON published a report from a Hamilton resident who claimed to have seen several strange, glowing orbs hanging over Lake Ontario. He also claimed to have seen these orbs on the lake several times prior, and what's more, he believes that there is an alien base under the water. He is far from the first person to have made such an assertion.

In August 1981, five witnesses who were driving alongside Lake Ontario early one evening saw a dome-shaped craft flying over the water. They followed the craft for some time before seeing it begin to descend and enter the water, disappearing from their sight. Perhaps the stories of an underwater alien base, stem from the 1977 book The Great Lakes Triangle by Jay Gourley, who made note that many planes and people had disappeared over Lake Ontario, not to mention the many UFO sightings of the area.

Unidentified Submerged Objects and Underwater Bases

Another book, Underground Alien Bases, released in 2012 and written by the somewhat strange "Commander X," has also perpetuated the legend of the alien base under the lake. It features within its pages several accounts of sightings on and around the lake and the assertion that an alien fortress of one kind or another lies under the water. However, none of the accounts can be verified by a secondary source and so are left open to debate as to how reliable they are.

Unidentified Submerged Objects and Underwater Bases

2.6 Flight Corridor To Underwater Base North Island, New Zealand

North Island in New Zealand has been witness to many strange objects entering the sea, with activity stretching out to some of the offshore islands around the area. UFO researchers state that the area is a hot spot for such activity, with locals claiming that the area is a flight corridor to an underwater base for the strange craft that exist there. Between January and March 1995, there were dozens of sightings of these objects coming and going in the area. These were not limited to nighttime, with many of the objects being witnessed during the daylight hours.

One particular daylight sighting on March 9, 1995, stands out. It began with two fishermen who witnessed a bright, silver, ball-shaped craft that seemed to glow or pulse as it moved and emitted a strange red stream behind it. The two men saw the object for less than 10 seconds before it vanished from their sight. Around two minutes later, however, there were sightings of what seemed to be the

Unidentified Submerged Objects and Underwater Bases

same object by the control towers at both Hamilton International and Rotorua Airports. Further supporting the incident were several reports that were phoned in to a local radio station from concerned residents, who all described a very similar object in the bright, sunny sky.

Unidentified Submerged Objects and Underwater Bases

2.7 Pacific Ocean 'Humming' Might be USOs

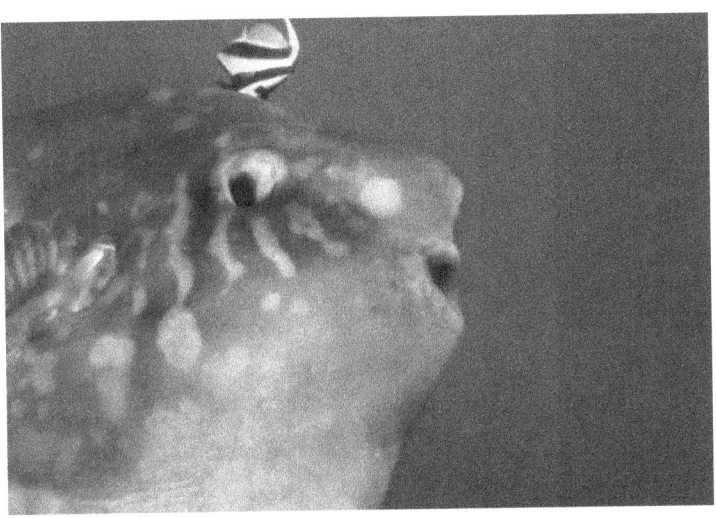

In the case of the alleged underwater alien base in the depths of the Pacific Ocean, the mainstream scientific explanation is actually stranger than that of the ufologists.

According to a research team led by Simon Baumann-Pickering, the low humming sound that has had people searching for bases and installations for over two decades is nothing more than sea creatures releasing gases from their swim bladders—essentially breaking wind. UFO researchers have almost entirely rejected this assertion.

The humming has been studied and debated since 1991 by scientists and UFO investigators alike, when it was first observed by National Oceanic and Atmospheric Administration (NOAA). Ufologists argue that rather than being a sound of the natural world, the sounds are more akin to those of electrically powered artificial structures. To them, this is proof of the existence of top secret

Unidentified Submerged Objects and Underwater Bases

underwater bases, which are quite probably extraterrestrial, given the advanced technology that would be required to build and operate such an underwater building.

Other theories include that the humming could be generated by marine vessels or an unknown geological phenomena.

2.8 Strange Lights and Sounds In Puerto Rico

The island of Puerto Rico has had numerous accounts of UFO sightings, particularly along its northeastern coastline, where strange craft have been seen both emerging out of and disappearing into the water for years. There is also a strong US Navy presence on the island, and stories from locals and US expats alike state that the US military at the very least is monitoring the strange activity on the waters along the coast.

Experienced ufologist and researcher Nick Redfern recently told of an ex–civil defense employee who had witnessed a strange craft emerge from the waters off the Puerto Rican coast. It steadily rose up into the air before shooting off at speed. Another account came to Redfern from a local police officer, who stated that the US Navy had spent considerable time tracking an underwater craft along the northeastern coast of the island.

Unidentified Submerged Objects and Underwater Bases

There have been numerous claims and theories that the immediate coastline of this Caribbean island is home to underwater alien bases, perhaps none more specific than those of investigative journalist Jorge Martin. Martin stated that using NOAA satellite images, they found several anomalies around the island which suggested artificial structures. Martin stated that these structures were unusually large and precisely rectangular and were generally on the eastern and southern coast of the island. He also claimed that coming off many of these structures, there appeared to be what looked like tunnels leading to other structures and even to the mainland.

What is interesting is that in the Puerto Rican city of Ponce during the late 1980s, local residents were complaining to authorities that they could hear loud rumbling, which seemed to be coming from beneath the ground. It was reported that the sounds were similar to those you might expect to hear when heavy machinery was operated. The sounds seemed to stop after several days. One of the places on the mainland that Martin's tunnels appeared to connect to was the city of Ponce.

Unidentified Submerged Objects and Underwater Bases

2.9 Ancient Alien Base at Lake Titicaca

The predominantly still waters of Lake Titicaca on the borders of Peru and Bolivia are not only the world's highest navigable waters, but also host to a plethora of UFO activity and according to some, home to an advanced and possibly ancient alien base.

The ancient city of Tiwanaku, considered to be one of the oldest cities in the world, sits on the southeastern shore of Lake Titicaca. Ufologists, ancient astronaut theorists in particular, state that the advanced level of agriculture, irrigation, and astronomy as well structures that suggest advanced building techniques leads them to believe that an extraterrestrial race once resided here. Numerous texts and statues have also been found in the immediate vicinity of the lake, which seem to depict ancient Mesopotamian underwater gods—gods that they claim were from an alien civilization that once resided under the still waters of Lake Titicaca.

Recently, a video surfaced on YouTube shot by a group of Italians who were on a spiritual journey of sorts in the region. As they pointed their camera out to the still, blue

Unidentified Submerged Objects and Underwater Bases

waters of Lake Titicaca, what appeared to be a large submerged object was seen slowly making its way away from the shore. It appeared to be disc-shaped object, and as it moved through the water, its shape remained the same, suggesting that it is indeed solid.

If there was an ancient alien underwater base below the surface of Lake Titicaca, perhaps this video may suggest that it is very much still in use.

Unidentified Submerged Objects and Underwater Bases

2.10 Underwater Alien Base at Guantanamo Bay

According to a former US Marine who served at Guantanamo Bay in the late 1960s, there is an underwater alien base off the Cuban coast. Furthermore, the Marine claimed that of the many UFO sightings of strange objects going in and out of the water, the US military has even managed to capture several on film. He also states that he and his colleagues were under strict instructions not to talk about the strange activity they witnessed there.

The Marine claimed that the craft he witnessed appeared to be made of a "dull" metal with a series of blue lights. When they would enter the water, the blue lights could still be witnessed but would grow fainter and fainter, suggesting to him that whatever the object was, it was descending deeper into the water.

Perhaps coincidentally, around 140 kilometers (90 mi) north of the US base in Gulf Breeze, Florida, there have been several UFO sightings that seem to match the descriptions of the craft as described by the unnamed marine, and may suggest they are originating from the same place.

One particular sighting came in November 1987, when Ed Walters claimed that he witnessed a strange object fly overhead from the coast. It emitted a bright, blue light that "trapped" him in its beam. He stated that while he was caught in it, everything around him was blue, and he could not move. Walters did manage to take several pictures of the craft. Hundreds of witnesses came forward to say they, too, had seen the strange object that evening.

There is debate over the credibility of his account. MUFON seemed to believe the sighting was genuine, while others have stated that the photographs were obvious fakes. In a strange twist, years after Walters' claims, an investigation into his story was said to have discovered a model craft very similar to the object that Walters stated he had photographed in a house where he once resided. Supporters of his claims, however, argue that this model is an obvious "plant" to discredit him.

2.11 Increased UFO Sightings at Lake Erie Cleveland, Ohio

Lake Erie, one of the five great lakes of North America, has a long history of strange lights and unexplained phenomena. However, the area around Cleveland, Ohio, has experienced an increase in UFO sightings over the years.

The 2011 book Eerie Erie: Tales of the Unexplained from Northwest Pennsylvania by Robin Swope examines and investigates numerous MUFON reports. Many of these reports speak of UFOs seen "crashing" into the water of Lake Erie. One particular reports from 1988 alleged a strange craft landed on the lake when it was iced over. The landing was witnessed by Sheila and Henry Baker, who made a report to the local Coast Guard. As the mysterious craft landed, there were strange sounds heard coming from the ice as well as a series of blue and red lights from the craft itself. There also appeared to be several strange triangular objects jettisoned from the descending object.

Unidentified Submerged Objects and Underwater Bases

These triangles seemed to move purposely around in all directions along the icy surface of the lake. Suddenly, the sounds on the ice stopped, and the craft and the mysterious triangles vanished, suggesting that they had indeed found their way below the ice and under the water.

According to a report in 2007 that ran on News Channel 5 in Cleveland, the area had seen no less than 20 UFO sightings in only two years. There are a great many videos on YouTube that claim to show some of these UFOs over Cleveland. Local UFO researcher Richard Lee stated that the UFOs seem to take particular interest in any new building projects undertaken in the area. Once a new project begins, according to Lee, there is usually a UFO sighting in the immediate vicinity not long after. Are they coming from their alleged base in Lake Erie?

Unidentified Submerged Objects and Underwater Bases

2.12 Lake Baikal—Scene of an Underwater Battle with Aliens

The ancient Lake Baikal in Siberia is said to be the deepest lake on the planet and has a history of UFO activity as well as alleged sightings of aliens underneath the water by Russian military divers.

These sightings date back to the Soviet era, but since the end of the Cold War, the reports have come into the public domain. Many of the accounts tell of a huge "mothership" hovering above the expanse of water and even of humanoid beings in strange, shiny suits climbing down from these ships and into the water.

Former Soviet naval officer and ufologist Vladimir Azhazha claimed to have leaked top secret files relating to an incident in 1982. He stated that military divers, who were on standard training exercises in the area, witnessed huge underwater craft that moved with a speed they had never seen before.

Several days later, the story took an even more bizarre twist when divers witnessed several strange beings under the water. They wore shiny suits and what appeared to be small, advanced oxygen masks. According to the alleged leaked documents, the unit was ordered to capture these strange creatures. When they attempted to do so, however, they were fought off with what appeared to be an advanced sonar wave weapon that ultimately killed three of the seven divers. The remaining four, now terrified and injured from their attempts detain the strange crew, retreated and made their report to superiors.
(Also see chapter 4.0 for more information about the swimmers.)

In 2009, photographs taken from aboard the International Space Station seemed to show two distinct saucer-shaped anomalies in the region, fueling further speculation that an extraterrestrial presence was under Lake Baikal and even the possibility that these "saucers" were the craft that had been witnessed nearly three decades earlier by the Soviet diving unit.

Unidentified Submerged Objects and Underwater Bases

3.0 Unidentified Submerged Objects

USOs are objects which either rise up or descend into the water. This can be in Oceans or large inland lakes. Some are near what we think are alien bases and others appear at unexpected locations around the world. In this chapter we have a variety of stories from the historic past and recent times about sightings of USOs.

3.1 The Utsuro Bune Story

In 1803, a round vessel drifted ashore on the Japanese coast and a beautiful woman emerged, wearing strange clothing and carrying a box. She was unable to communicate with the locals, and her craft was marked with mysterious writing. This story of an *utsurobune*, or "hollow ship," in the province of Hitachi (now Ibaraki Prefecture) is found in many records of the Edo period (1603–1868), and Tanaka Kazuo, professor emeritus at Gifu University, has studied the topic for many years. What drew him away from his main research area, applied optics, to investigate this curious episode? And what really took place?

"Like a Flying Saucer"

Tanaka says he began to research the ship after the deadly subway sarin attacks in 1995 by the Aum Shinrikyō cult. "There was a lot of coverage of Aum founder Asahara Shōkō's prophecies and claims to be able to float in the air. Yet the cult's senior members were part of the scientific elite. I started giving lectures considering paranormal phenomena from a scientific perspective, which meant that I was collecting all kinds of materials for teaching, such as about UFOs in the United States and

Unidentified Submerged Objects and Underwater Bases

Japanese folklore. While doing so, I came across the *utsurobune* legend." He adds, "Long before the American UFO stories, the craft depicted in Edo-period Japanese documents for some reason looked like a flying saucer. This was fascinating to me."

UFOs became a modern sensation after the media reported US businessman Kenneth Arnold as having witnessed "flying saucers" on June 24, 1947. A flood of similar stories followed from around the world. Most famously, a UFO was alleged to have crashed to the ground near Roswell, New Mexico, in July 1947. "In the end, though, no wrecks or alien bodies were recovered," Tanaka says. "There was only the ambiguous testament of witnesses. It was the same with all the other UFO stories from around the world—they were mysteries without any substantial evidence. The *utsurobune* legend, however, has a number of documents to examine as leads, so in this sense, for researchers it's a mystery with substance."

A Ninja's Report

There are similar oral traditions about "hollow ships" across Japan in the Edo period. Tanaka's research is focused on the various documents that describe the 1803 incident in Hitachi and include illustrations of a beautiful woman and a strange vessel, although they cite different dates. One of the best-known sources is the *Toen shōsetsu* (Toen Stories), an 1825 collection recording fantastic rumors, which was written by the Toenkai literary circle and edited by Kyokutei Bakin, famous for his lengthy historical romance *Nansō Satomi hakkenden* (The Eight Dog Chronicles). Others are Nagahashi Matajirō's 1844 work *Ume no chiri* (Plum Dust), as well as collections like *Ōshuku zakki* (Ōshuku Notes), *Hirokata zuihitsu* (Essays by Hirokata), and *Hyōryūki-shū* (Records

Unidentified Submerged Objects and Underwater Bases

of Castaways), which gathers stories of foreign ships washed up in Japan and of Japanese sailors who came ashore overseas.

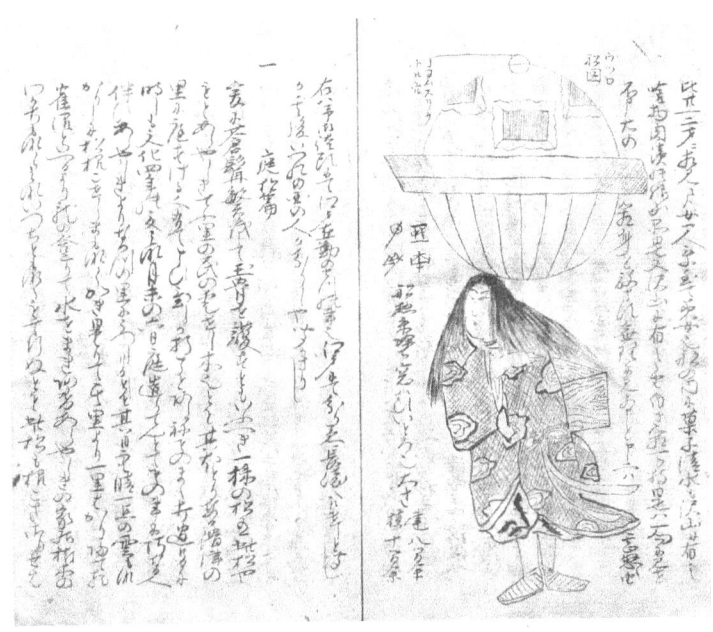

From Ōshuku zakki (Ōshuku Notes; around 1815) by Komai Norimura, a vassal of the powerful daimyō Matsudaira Sadanobu.

Unidentified Submerged Objects and Underwater Bases

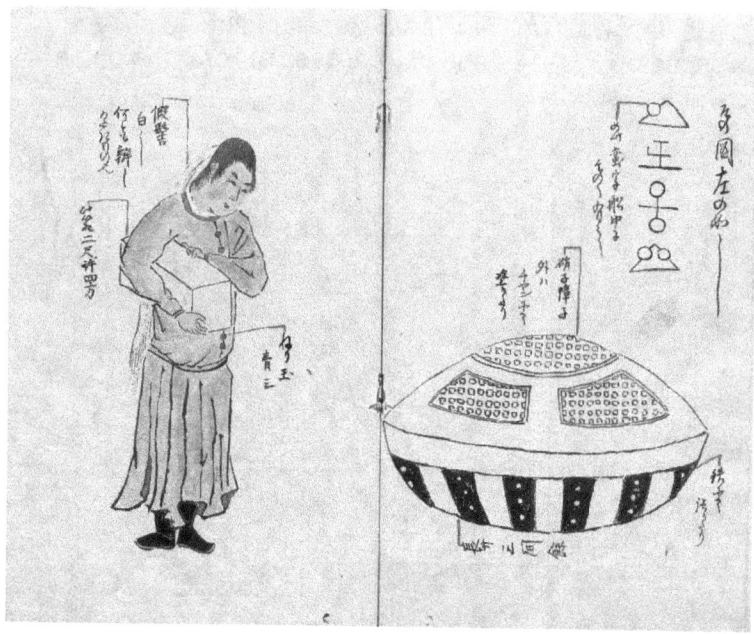

From Hirokata zuihitsu (Essays by Hirokata; 1825) by shogunate retainer and calligrapher Yashiro Hirokata, who was also a member of the Toenkai circle.

At first, Tanaka theorized that the incident was an embellished account concerning a shipwrecked Russian whaler, but he could not find any mention of such a disaster in official records. Instead, he discovered new materials, becoming absorbed in further background research. To date, he has found 11 documents relating to the Hitachi *utsurobune* legend, of which the most interesting are thought to date from 1803, the same year that the craft was said to have come to shore.

One is the *Mito bunsho* (Mito Document) owned by a collector in Mito, Ibaraki Prefecture. Tanaka noticed that the woman's clothing in an illustration in the work was similar to that of a bodhisattva statue at the Shōfukuji temple in Kamisu, also in Ibaraki, which is dedicated to the raising of silkworms. A legend credits the start of sericulture in the area to a Princess Konjiki (or "golden princess"), who is a motif in images at the temple. In one version of the story, Princess Konjiki is washed up to shore after traveling from India on a dugout boat in the shape of a cocoon. She repays the kindness of a local couple who try to nurse her back to health by bestowing on them the secrets of sericulture when she herself becomes a silkworm after her death. Among the various materials, only the illustration in the Mito document appeared greatly similar to Princess Konjiki. Tanaka thinks that when the first rumors of a "hollow ship" coming ashore at a beach called Kashimanada were spreading, the people at Shōfukuji may have decided to incorporate it into promotion of the temple.

Another even more important source is the *Banke bunsho* (Banke Document) owned by Kawakami Jin'ichi, the heir to the Kōka *ninjutsu* (ninja arts) tradition and a *ninjutsu* researcher and martial artist. It is named after the Banke, or Ban family, of Kōka ninja. While some other materials say the vessel came ashore at locations like Harayadori or Haratonohama, there is no evidence that such places exist. This document, however, records the location as Hitachihara Sharihama, which appeared in a map produced by the famous cartographer Inō Tadataka, and is now known as Hasaki Sharihama in Kamisu.

Tanaka comments that while the other materials show geographical inconsistencies, this document mentions a real place name. He says that Kawakami suggested a Banke member might have been assembling information while working for the head of the Owari domain (now Aichi Prefecture). "If so, he wouldn't record any untruths, so we can say the document is very reliable."

Unidentified Submerged Objects and Underwater Bases

Unidentified Submerged Objects and Underwater Bases

3.2 Christopher Columbus and the light sightings

Christopher Columbus and his small fleet of three ships, the Nina, the Pinta, and the Santa Maria, are credited with discovering the "New World". Few people, however, are aware of the very strange events that occurred during their voyage. These events were recorded in the ship's log.

On the 11th of October 1492, at approximately 10pm, Columbus and his crew were sailing across one of the deepest ravines in the Atlantic, almost four miles deep, and through what is today known as the Bermuda Triangle.

Pedro Gutierrez was a crew member of the Santa Maria, sailing with Columbus. He noticed a strange light shining out in the distance over the sea.

What was so odd about this sighting, making it particularly intriguing, is that the light was observed coming up from the water. The fact that it appeared at least four times and was seen in a variety of positions rules out many logical explanations such as a comet, meteor, bright planet or star.

Unidentified Submerged Objects and Underwater Bases

The initial sighting was allegedly followed by a great flash of light with a level of brilliance unlike anything these men had previously known. The light is described as suddenly erupting in the sky, startling Columbus and his crew - and the crews of the other two Spanish ships.

This event occurred only five hours before Columbus and his men would discover the New World.

It was Columbus himself who maintained the ship's log. This rare and valuable handwritten document is today held by the Fordham University in New York. Archivists at this institution have made the logs contents available to the general public.

Columbus described the light as having the appearance of a flickering wax candle going up and down in the night. The light could not have been caused by a camp fire on land because of its initial, dazzling brilliance and also the fact that land would have been well beyond the horizon from the position of the Spanish ships. From reading the logs, the 11th of October sighting was not an isolated occurrence.

Over the two-month journey, Columbus's log exhibits a consistent pattern of peculiar and cryptically reported incidents, including unexplained sightings and unusual events witnessed occurring in the skies.

On the 10th of September 1492, around about the halfway point of the voyage, the crew of the Nina stated they had seen two different species of bird, both of which are known never to stray further than 25 leagues from the land.

On the 11th of September 1492, the log recorded that the fleet had happened upon the floating mast of a ship

Unidentified Submerged Objects and Underwater Bases

estimated to be 120 tons in weight and so large that they were unable to retrieve it for closer inspection.

On the 17th and 20th of September, the logs tells us that the crew witnessed multiple bright lights in the sky that were seen to move in relation to the stars.

Some experts have suggested that Columbus played these events down when recording them in the ships log. The reason they give is that he might have feared encouraging accusations of insanity among members of the crew who had never before witnessed such strange events.

Superstition and belief in bad omens were commonplace back in the 1490s and Columbus might have feared being locked in the brig on suspicion of insanity or the practice of black magic. He might also have feared similar accusations or ridicule upon his eventual return to Spain.

Few doubt the written word or the integrity of Columbus over these events but more recent analysis has led to numerous suggestions as to their cause.

Unidentified Submerged Objects and Underwater Bases

Unidentified Submerged Objects and Underwater Bases

3.3 Puerto Rico Sightings

UFO SPLITS INTO TWO AND SPLASHES OVER WATER, AMAZING NEW FOOTAGE , Leaked US Security Footage Of a Confirmed UFO Sighting, The object in the video1 appears to look like a metallic sphere, which moves fairly quickly over land and then into the ocean and seems to tumble or change shape, almost like the UFO is 'morphing' into something different. The SCU have confirmed that the leaked footage has come from a DHS whistle-blower who does not wish to be named. Apparently since the video1 has been released by the SCU, quite a few other witnesses have come forward to confirm the sighting. The SCU apparently spent two years reviewing the footage and building up an in-depth report to confirm the videos authenticity. It has been confirmed that at least scientists and researchers have been named against this project.

Unidentified Submerged Objects and Underwater Bases

This UFO incident allegedly started at about 9.20 pm on April 25 2013 at the Rafael Hernandez Airport in Aguadilla, Puerto Rico and was "monitored by the crew of a DHC-8 Turboprop aircraft from US Customs and Border Protection (CBP) – who are part of the
US DHS.

Recent 'leaked' video1 of what appears to be a UFO coming out of the sea and apparently splitting in two is alleged to belong to the US Department of Homeland Security (DHS) and has got truth seekers all over the world stating that this is the real deal and could not possibly be anything else. The video1 alleged to have been leaked from the US Department of Homeland Security (DHS) shows a "UFO being tracked by thermal imaging camera" recently appeared online.

The almost 4 minute long video1 is being taken seriously by UFO and alien researchers all over the world and was placed online by a group calling itself the Scientific Coalition for Ufology (SCU). According to the SCU website, its researchers confirm that the UFO "exhibits characteristics that cannot be explained by any known aircraft or natural phenomenon".

This past weekend, former U.S. Navy Commander David Fravor was a guest on the Joe Rogan Experience podcast. Fravor, who was the subject of a New York Times article about his 2004 UFO sighting, discussed a spooky new sighting a fellow pilot revealed to him after they were both out of the Navy.

According to Fravor, the eyewitness was a former pilot of the MH-53E Sea Dragon, the Navy version of the Marine Corps' CH-53E Sea Stallion, based at Naval Station

Unidentified Submerged Objects and Underwater Bases

Roosevelt Roads, on the island of Puerto Rico. Twice while recovering spent practice munitions out of the water, the pilot spotted a weird underwater object.

In the first incident, the pilot saw a "dark mass" underwater as he and his team retrieved a flying practice drone. The pilot described the object as a "big" mass, "kinda circular," and he was certain it wasn't a submarine. In the pilot's second sighting, a practice torpedo that the pilot was sent to recover was "sucked down" into the depths of the ocean in the presence of a similar underwater object. The torpedo was never seen again.

Elsewhere in the interview, Fravor reveals that a 79-year-old woman contacted him after his sighting went public. The woman explained that her father, a naval officer, was at one time based at the naval station in San Francisco in the 1950s. When she was a child, her father showed her a telegram that stated unidentified objects had been sighted going in and out of the water at a now forgotten set of latitude and longitude coordinates. The woman's father told her, "We get these all the time, and it's always in the same area."

Unidentified Submerged Objects and Underwater Bases

Unidentified Submerged Objects and Underwater Bases

3.4 Japanese Fishing Boat

In 1970, biologist Ivan Sanderson published the book "Invisible Residents". Sanderson, a noted student of unusual phenomena, devoted the book to sightings of what were later called Unidentified Submerged Objects, or USOs. USOs are defined as unknown craft that are sighted in the water, sighted rising up out of the water, or diving into the water. Sanderson catalogued scores of reports of USOs:

On the 19th of April, 1957, crew members aboard the Kitsukawa Maru, a Japanese fishing boat, spotted two metallic silvery objects descending from the sky into the sea. The objects, estimated to be ten meters long, were without wings of any kind. As the hit the water, they created a violent turbulence.

The exact location was reported as 31° 15' N and 143° 30' E.

Unidentified Submerged Objects and Underwater Bases

Unidentified Submerged Objects and Underwater Bases

3.5 Another Sighting off Puerto Rico

Sanderson also reports an incident that reportedly took place off the coast of Puerto Rico in 1963 during an anti-submarine warfare exercise.

The maneuvers were conducted off Puerto Rico in the Atlantic some 500 miles southeast of the continental United States. All reports seem to agree that there were five "small" naval vessels concerned, but in more than one account the aircraft carrier Wasp is stated to have been the command ship…

A sonar operator on one of the small vessels, otherwise listed as a destroyer, reported to his bridge that one of the submarines had broken formation and gone off in what appeared to be pursuit of some unknown object. This operator did not, of course, know if this was a "plant", since the maneuvers they were engaged in were exercises designed to train personnel in detection of enemy craft...However, this operator's report was not all within the limits of any such simulation,. Trouble was that said subaqueous object was traveling at "over 150 knots"! Was the vehicle a, Warship, Heavy cruiser, Battleship, Naval ship, Boat, Ship, Aircraft carrier, Destroyer, or Destroyer escort?

According to Sanderson, "no less than craft," including anti-submarine warfare patrol aircraft, tracked the high-speed, unknown object. Furthermore:

It is said that technicians kept track of this object for four days, and that it maneuvered round about, and to depths of 27,000 feet.

3.6 Half Moon Bay, California

Off the coast of Half Moon Bay, California, an eyewitness reported that in 2007 she observed three UFOs while aboard the cruise ship Dawn Princess (renamed in 2017 Pacific Explorer.)

"After about 5 minutes, three softly glowing objects came into view – three uniform, nearly spherical objects, evenly spaced in a line parallel to the ship's hull and hovering just above the water surface… They appeared to stay in one place while the ship moved past them. They were hovering, but didn't disturb the water below them. Just as they went out of sight, the left one (toward the bow) splashed down into the water and disappeared."

Unidentified Submerged Objects and Underwater Bases

Unidentified Submerged Objects and Underwater Bases

3.7 Imperial Beach, California

One report logged in April 2019 states that an object resembling a "small white boat" flew up out of the water near Imperial Beach, California, "at about [500] feet." The object promptly "flew south at a very high rate of speed."

Whatever USOs are—figments of the imagination, mechanical malfunctions, secret government craft, or even the work of extraterrestrials—there's a long history of sightings. Fravor's anonymous helicopter pilot is just the latest in a long line of mysteries, and they won't be solved anytime soon.

Unidentified Submerged Objects and Underwater Bases

Unidentified Submerged Objects and Underwater Bases

3.8 Nuclear Submarine Stories

"Marc, who runs a special effects company called FX Models that undertakes Naval contracts, said: "As a thank you for doing some work for them Navy asked me if I wanted to go for a ride in a submarine so I said yes.

"Once we got under I was sitting in the sonar station and the sonar operator was sitting right next to me.

"Submarines are loud – people think they are very quiet and it's true they are on the outside because the sound doesn't get out. But inside you hear fans, noise - it's a constant din on a sub.

"I was sitting there zoning out a little because I was sea sick and all of a sudden the sonar kid shouts 'fast mover, fast mover' and I'm jolted awake – thinking 'What's happening? Is it a torpedo?'

Unidentified Submerged Objects and Underwater Bases

"The executive officer comes out and the operator shows him the path of the object and the officer says 'How fast is that going?'

"And the kid said 'several hundred knots'. I start to lean forward to listen in – and the officer said 'Can you confirm it?'

"So he goes to another sonar machine and confirmed it wasn't a machine anomaly - it was real. I thought 'Wow that is incredible'.

"When the sonar guy said 'What do I do with this?' the officer said 'log it and dog it' - in other words log it and bury it."

Four years later Marc said he was doing some more contract work for the Navy when he spoke to a senior naval figure about what he saw.

"I asked him 'Can you tell me about the Fast Mover Program?'" Marc explained.

"He looked at me and said 'Sorry Marc I can't talk about that program'.

"So he basically confirmed to me that the program exists - he said everything without saying anything.

"What that told me was that USOs are common – we even have a program in place to classify and log and determine the speed of them and it goes into a vault."

Marc made the claims at the Devil's Tower UFO Rendezvous in Hulett, Wyoming – where UFO enthusiasts

Unidentified Submerged Objects and Underwater Bases from around America met at the site of Close Encounters of The Third Kind rock to discuss alien-related findings.

Unidentified Submerged Objects and Underwater Bases

Unidentified Submerged Objects and Underwater Bases

3.9 Lake Baikal, Russia

(See Chapter 2.11 for additional stories on Lake Baikal)

Curious reports of extra-terrestrials are pulling Steven Spielberg to Siberian jewel Baikal.

True or not about Spielberg's interest, the lake is perhaps the biggest focus of UFOs in Russia.

Since ancient times, the vast Lake Baikal has been known as deeply mysterious, but in the closing years of the Soviet era, and since, it has been the location of a number of alleged sightings of aliens and UFOs.

Initially these were covered-up by the authorities of the USSR, but later they were revealed by the Russian media.

In recent days there have been unconfirmed reports in Russia that American director Steven Spielberg is planning a documentary based on these weird and unexplained accounts. At the time of writing, this appeared to be a hoax,

Unidentified Submerged Objects and Underwater Bases

though it was unclear who planted stories in the Russian media.

True or not about Spielberg's interest, the lake is perhaps the biggest focus of UFOs in Russia.

The versions of extra-terrestrial activity at Baikal - edged by mountains and containing one-fifth of the world's unfrozen freshwater - relate to supposed aliens seen by military divers in its depth, and large 'spaceships' hovering over its grey, moody expanses.

Some of the images here show what two photographers claimed were UFOs buzzing the lake, while others are mock-ups from NTV based on descriptions of an incident at Kudara-Somon, in Buryatia, exactly a quarter of a century ago.
A number of sightings also indicate bright 'cigar-shaped' objects in the sky flying over Baikal, as in the top picture.

NTV channel mock-up of 'flying saucer' based on accounts by residents of Kudara-Somon village in 1990.

Unidentified Submerged Objects and Underwater Bases

A case for which there were no images, but an intriguing description, occurred at Kurma, Irkutsk region, in 17 April 1987. The words that follow are from Valery Rudentsov, a local resident of nearby Shida village:

'There were 13 of us. At about 12.20am, one of our guys went out into the yard, a few seconds later runs, and calls all of us out. He stood in the center of the yard and pointed his finger at the sky.

'Diagonally from his gesture - 150 metres above us - hung a huge flying saucer. From the center of the plate went a phosphorescent purple ray. And at the edges of the plate were yellow portholes, almost like in our rural houses. The diameter of the plate was 70 metres. We saw it so clearly and for a long time, someone even suggested he throw a stone at it ...

'The weather was amazingly quiet, no sound was heard from the hanging saucer, although behind us was the village of Kurma - there was the noise of a dog barking, the lowing of cows. We were spellbound.

'It was a full moon and the visibility was so clear that no one of us could doubt the reality of what we saw. And then the plate slid smoothly away, sailed along the shore of the bay and further slipped into the hills of Olkhon. Neither before nor after have I ever met such a thing.

'But since that time it has been a kind of sacrilege to me - not to believe in UFOs. My friend Alexander, a hunter, and his colleagues who lived there for 20 years, often see UFOs - and all is fine, he is still alive. So if to speak about glowing balls or 'cigars', we constantly see these on the shores of Lake Baikal. They exist.'

Unidentified Submerged Objects and Underwater Bases

The case highlighted by NTV channel was on 16 May 1990 in the village of Kudara-Somon, in Kyakhtinsky district, some 300 km from Ulan-Ude, capital of the Republic of Buryatia. Olga Fedorova, a local resident, recalled: 'At some moment everything turned yellow. My daughter came home from school. I looked - her face was yellow.'

The explanation soon became clear, according to accounts from villagers.

Vasily Timofeev spoke of a flying saucer. 'Its diameter was around 30 meters, it shone brightly. But I did not see a clear image of metal or something like this.'

Another resident Margarita Tsybikova said: 'From this dish came down people in shiny, shimmering costumes.' Olga explained: 'There were people, as far as I remember, three people in shining yellow suits. Seems there were people, yes.' Marina Zimireva, who also says she witnessed this extraordinary sight, said: 'It was some kind of circle, it can be said, it was like a disk. It turned on the edge and, well, windows were visible.

'I personally decided for myself that they were people. They had some human image. They were the same - straight, slender, they had arms and legs. And their gait was the same as ours. A little lower down there were three in orange suits. They went down from the disk like a man - the steps were very visible.'

Then, as they recounted the strange event, the 'aliens' saw the people watching them. They returned to their spaceship and flew away.

Unidentified Submerged Objects and Underwater Bases

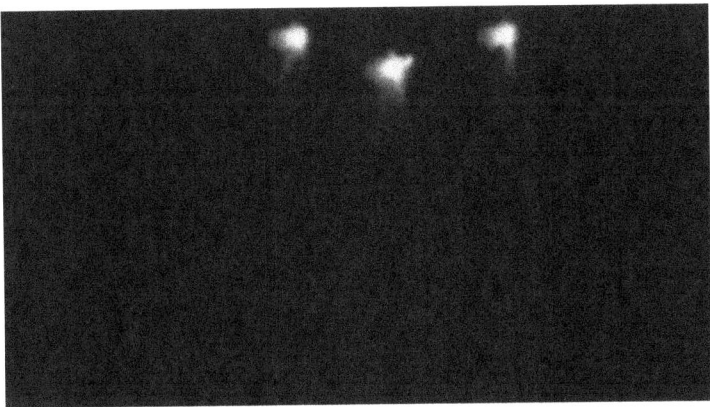

A picture taken by Nikita Tomin, shows three green-shaded lights on a UFO flying above a lakeside resort in Irkutsk region.

Game warden Gennady Lipinsky also recalls seeing a UFO. 'When I saw it, it was flying low. Until it disappeared over the horizon, I kept looking at it. I call it a fireball, and what it really was - I cannot know.'

The chairman of the Union of Photographers of Buryatia, Sergey Konechnykh, Ulan-Ude, was quoted about a much more recent incident, on 9 July 2009, at around 10pm. 'My son and I went out to the balcony, to see the last of the waning sunset. Suddenly there appeared these two glowing points and they hovered over the water.'

His pictures of this incident are clear yet perhaps raise as many questions as they answer. They show two glows in the night sky, featuring a yellow core with an orange-red surround.

By his account, the mysterious crafts rose and rebased elsewhere on Baikal's surface.

Around the same time, Anna Vinogradova, recalled a different but equally strange sight on the water. 'We were standing with

Unidentified Submerged Objects and Underwater Bases

tents on the beach and at night we saw orange-red flashes, as if a huge fires,' she said.

At Listvyanka in July 2010, a picture taken by Nikita Tomin, shows three green-shaded lights on a UFO flying above the lakeside resort in Irkutsk region.

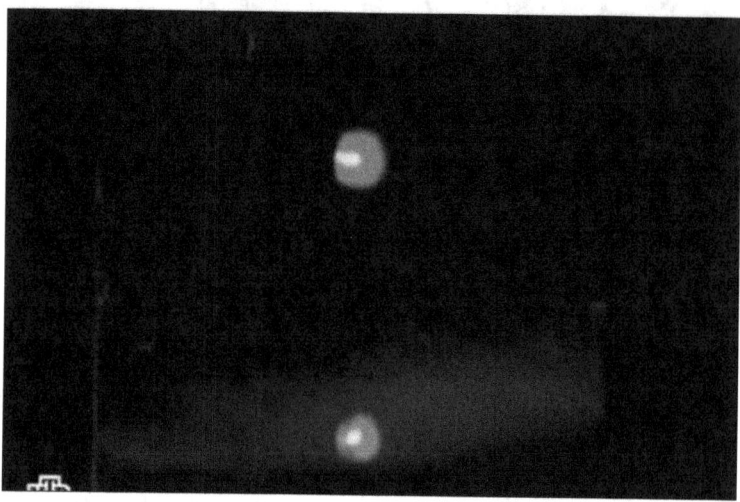

The chairman of the Union of Photographers of Buryatia, Sergey Konechnykh shows the pictures he made in July 2009.

'It flew right above us, very low. The object was shining down on us with a green light. We were a bit scared,' he said.
But the accounts also include 'aliens' in the deep waters of Baikal. Interestingly, unlike several other Siberian lakes, Baikal is not seen as home to a Loch Ness-style monster, but rather boasts space-like aliens under its murky surface.

For example, Vyacheslav Lavretevich, a rescuer, recalled an incident but did not give the date. 'We were on a yacht on Lake Baikal, and from under us flew out a huge glowing disk. It blinded us, and for a second flew into the sky.

Unidentified Submerged Objects and Underwater Bases

'We did not even have time to grab any cameras, nor take video, although many of us saw it. It was a huge - and lit up all of our yacht. In diameter it was probably 500-700 metres, a huge disk.

'For three minutes it shone from below (the surface), and then abruptly departed in a second. The spectacle was huge, awesome. Beautiful, and shocking.'

Oleg Chichulin was also on a boat, training students. Near Cape Svyatoi Nos they saw strange objects.

'There was a ball that glowed. And then this ball started to fade, fade and blush. And it turned into a red ball. This red ball for a while lay on the water, and then began to sink. And all this gradually, gradually went under the water. And it became dark.'

Yet there are even more intriguing accounts of underwater aliens in the vast lake.

In 1977, when Leonid Brezhnev ruled the Kremlin, two researchers named V Alexandrov and G Seliverstov, were in a submersible device at a depth of 1200 metres in the lake. The researchers turned off their spotlights, to explore the depth of penetration of sunlight into the water. Suddenly the scientists were bathed in light from an unusual 'glow'.

Aleksandrov recalled: 'It was so like if our device was lit from above and the side by two strong spotlights. Only a minute later, unknown floodlights went out, and we found ourselves in total darkness.'

Unidentified Submerged Objects and Underwater Bases

In 1977 two researchers named V Alexandrov and G Seliverstov, were in a submersible device at a depth of 1200 metres in the lake and observed strange lights.

In 1982, seven military divers were reported to have come across aliens under the waters of Baikal. Alexey Tivanenko, a doctor of history, said: 'At a depth of 50 metres, they met swimmers, around three metres tall, dressed in tight-fitting silvery suits. They did not have any scuba or other devices, just helmets on their heads. (See the full Swimmers story later in this book.)

'They received an order to catch the Ihtiander (half-boy, half-shark, from modern Russian folklore) - but they were immediately washed ashore with signs of decompression. 'They had two decompression devices, but one was broken. All seven people could not be put inside, so they put only four of them. And those three people, who were not put in the device, died on the beach.'

Unidentified Submerged Objects and Underwater Bases

Tivanenko has likened the descriptions to ancient petroglyphs seen by some as being aliens visiting Earth. 'I have hundreds of drawings with these 'Sons of the Sky',' he told NTV.

'They are united by the fact that all of them are tall, dressed in suits, all with the helmets on their heads. And there are mechanisms used by astronauts today.'

Unidentified Submerged Objects and Underwater Bases

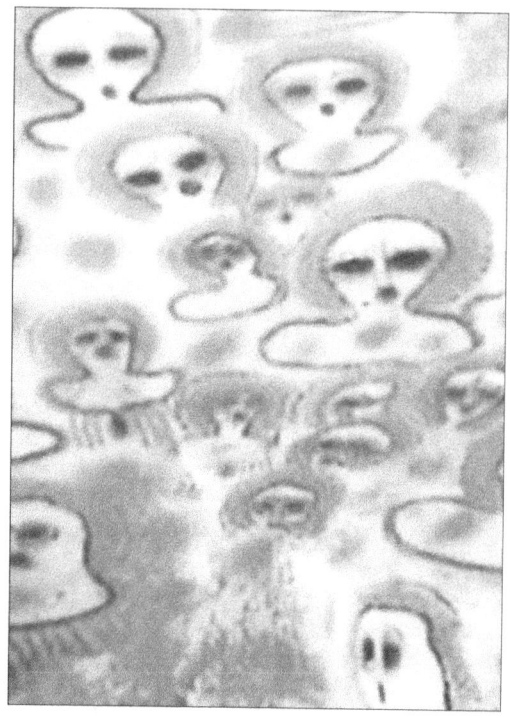

'I have hundreds of drawings with these 'Sons of the Sky', told Alexey Tivanenko.

Reports surfaced several days ago in the Russian media that Spielberg was expected at Baikal in May, and that he intends to make a documentary entitled Depth 211.

Citing the 'press service' of MUFON - the Mutual UFO Network, a US-based organization that investigates UFO sightings - news source <u>infobaikal.ru</u> and others reported his trip.

Yet there was no confirmation from the Hollywood director, who has made such films as Close Encounters of the Third Kind and E.T. the Extra-Terrestrial, would be arriving, nor further detail on the reported project.

Unidentified Submerged Objects and Underwater Bases

Later, Komsomolskaya Pravda and IA Teleinform denied the reports but without quoting any sources close to the director.

Unidentified Submerged Objects and Underwater Bases

Unidentified Submerged Objects and Underwater Bases

3.10 Russian Subs and USO Battle

Russian submarines in secret battle with 'ALIENS' deep under the oceans, top secret Kremlin documents claim.

RUSSIAN submarines are fighting a secret war with "alien" craft deep under the oceans, according to top secret Kremlin documents.

In scenes straight out of a Hollywood movie, Soviet subs are playing a game of cat and mouse with strange underwater craft, according to a new book called Russia's USO Secrets, by Brit investigator Philip Mantle, based on Russian documents and accounts from military veterans. USOs (underwater submersible objects) are the UFOs of the oceans

Describing a bizarre incident in the Bermuda Triangle in 2009, former nuclear submarine commander Yury Beketov said: "We repeatedly observed that the instruments detected the movements of material objects at unimaginable speed, around 230 knots (400 km per hour).
"It's hard to reach that speed on the surface – only in the air [is it possible]… The beings that created those material objects significantly exceed us in development."

The book also contains accounts from Russian veterans about their close encounters - including hostile acts - at sea during the Cold War.

They have described how UFOs would apparently appear in the sky before diving down beneath the waves.

Unidentified Submerged Objects and Underwater Bases

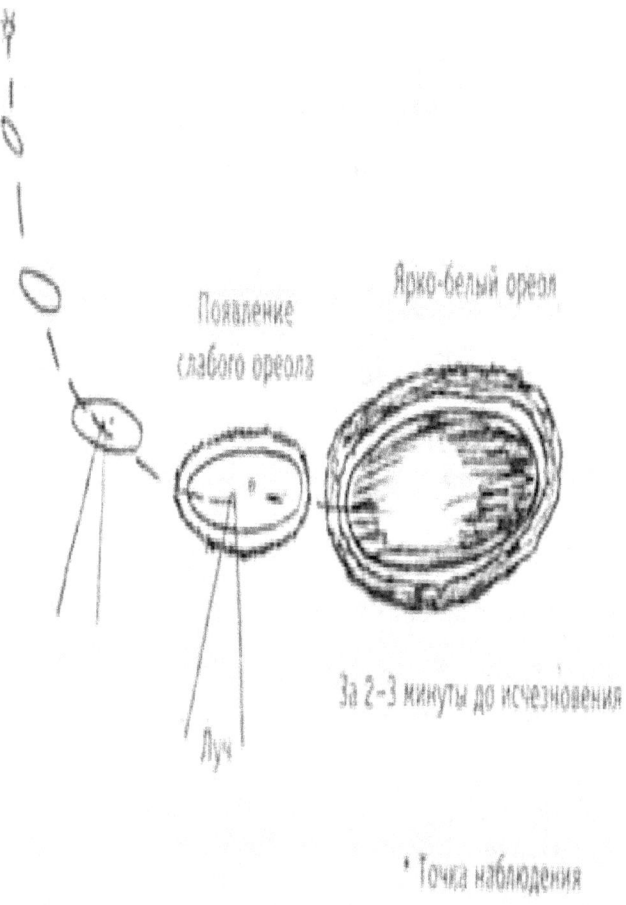

Рис. 57. Схема полета объекта, нарисованная К. М. Меджидовым

In one terrifying incident, a captain ordered depth charges to be dropped into the path of a USO which changed direction and sped off.

Unidentified Submerged Objects and Underwater Bases

Lieutenant-Commander Oleg Sokolov told students that while on duty he spotted a strange object through a periscope rising out of the water.

The book also reveals how another crew watched as a cigar shaped UFO shot three beams of light down into the Atlantic Ocean.

The UFO was about 200-250 metres long and slowly descended into the ocean making a strange hissing sound about half a mile from the submarine. But the object was not tracked on radar.

Marine scientists in the port city of Sevastopol, Ukraine, claim they spotted a huge "wheel" rotating below the Black Sea while they were deep sea diving.

Picture taken in Kazakhstan in April 1995

Unidentified Submerged Objects and Underwater Bases

```
В приложении содержится информация о случаях наблю-    Стр.
дения аномальных явлений в районах:
- города Петропавловска          20 октября 1982 г.   -.....3
- городов Курска,Воронежа,Ельца  17 октября 1983 г.   -.....24
- населённого пункта Гюзен       3 февраля  1985 г.   -.....34
- Хабаровского края              23 мая     1985 г.   -.....39
- Приморского края               12 ноября  1985 г.   -.....42
- города Магадана                25 ноября  1986 г.   -.....50
- полуострова Тикси              14 августа 1987 г.   -.....56
- города Минеральные Воды        14 декабря 1987 г.   -.....60
- города Невинномысовска         30 декабря 1987 г.   -.....73
- Камчатской области                       1987 - 1988 г.г.-.....75
- города Хабаровска              6 мая      1988 г.   -.....78
- города Магадана                1 октября  1988 г.   -.....85
- города Сочи                    26 июля    1989 г.   -.....87
- города Капустина Яра           28 июля    1989 г.   -.....92
- Астраханской области           28 сентября1989 г.   -....111
- Магаданской области            21 октября 1989 г.   -....120
- Владимирской области           21 марта   1990 г.   -....123
```

A list of cases from the KGB file release

In a 1951 incident, documented in records, a Soviet submarine encountered a gigantic underwater object heading towards the shores.

The captain ordered depth bombs to be dropped into the path of the USO but the object did not react to the attack and stayed on its course before darting towards the surface. At a depth of 50 meters it stopped its ascent, changed course and departed.

Russian sailors again observed a UFO in the Mediterranean in July 1978, according to the book.

The captain of Soviet motor ship Yargora immediately sent a radiogram to the Soviet Academy of Sciences in Moscow telling them the object was shaped like a flattened-out sphere and was a white pearl color, files reveal.

Unidentified Submerged Objects and Underwater Bases

3.11 UFO Crash into Water over Hawaii

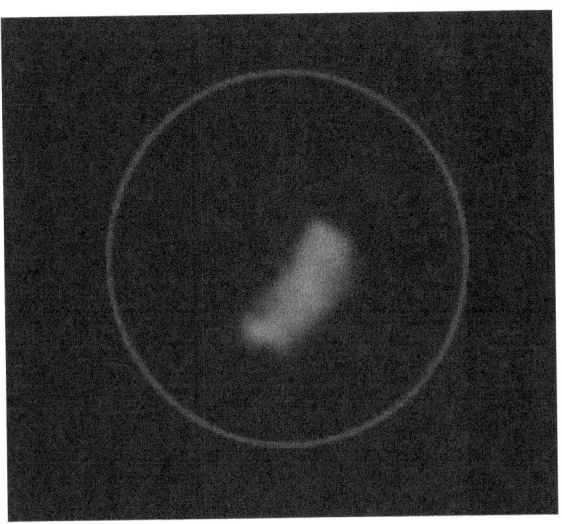

Video of blue 'UFO' crashing into the ocean near Hawaii makes waves on internet

Multiple eyewitnesses alerted the police and aviation regulators after they spotted a bright blue object over the skies in Oahu on January 5.

Pictures and videos of the strange object on social media showed a bright blue mass, moving across the night sky.

Residents in Hawaii were left baffled after they spotted an Unidentified Flying Object (UFO) in the sky and falling into the ocean, which prompted them to notify the police.

According to local news reports, multiple eyewitnesses alerted the police and the Federal Aviation Administration

Unidentified Submerged Objects and Underwater Bases

(FAA) after they spotted a bright blue object over the skies in Oahu on January 5 at around 8.30 pm.

Pictures and videos of the strange object on social media showed a bright blue mass, moving across the night sky.

Unidentified Submerged Objects and Underwater Bases

3.12 USOs at Guantanamo Base, Cuba

Marine Claims UFOs 'Flew out Of Underwater Base near Guantanamo Bay'-March 10, 2016

A former U.S. Marine has made bizarre claims that he and his comrades saw dozens of huge, hi-tech spacecraft flying out of an 'underwater alien base' next to the Guantanamo Naval Base in the late Sixties.

The Marine - who has chosen to remain unnamed - claims that he was forbidden to talk about the shimmering 100ft craft he had seen, and claims he saw intelligence officers filming the craft.

The interview comes courtesy of UFO researchers MUFON - and thus should be taken with a pinch of salt, as the site is known for printing dubious accounts of alien encounters uncritically.

The Marine said, 'All of us marines were amazed at the amount of UFO activity over and around this base.

Unidentified Submerged Objects and Underwater Bases

'Virtually every night UFOs were flying overhead with altitudes of less than 300 feet.'

'When I stood guard duty on the south side of the base, I witnessed on many, many nights UFOs landing and taking off out of the ocean.'

'There were large blue lights moving around after their landing in the ocean and then slowly dimming down as they obviously descended deeper.'

The unnamed soldier said that the encounters occurred in the late 60s - between 1968 and 1969.

He says that the machines he saw were not flying saucers, describing one as a 'huge white cloud with a blue/white, baby blue pulsating light in the middle of it.'

Another Report from Guantanamo:

The Marine claimed that the craft he witnessed appeared to be made of a "dull" metal with a series of blue lights. When they would enter the water, the blue lights could still be witnessed but would grow fainter and fainter, suggesting to him that whatever the object was, it was descending deeper into the water

Perhaps coincidentally, around 140 kilometers (90 mi) north of the US base in Gulf Breeze, Florida, there have been several UFO sightings that seem to match the descriptions of the craft as described by the unnamed marine, and may suggest they are originating from the same place.

One particular sighting came in November 1987, when Ed Walters claimed that he witnessed a strange object fly overhead from the coast. It emitted a bright, blue light that "trapped" him in its beam. He stated that while he was caught in it, everything around him was blue, and he could not move. Walters did manage to take several pictures of the craft. Hundreds of witnesses came forward to say they, too, had seen the strange object that evening.

There is debate over the credibility of his account. MUFON seemed to believe the sighting was genuine, while others have stated that the photographs were obvious fakes. In a strange twist, years after Walters' claims, an investigation into his story was said to have discovered a model craft very similar to the object that Walters stated he had photographed in a house where he once resided.

Supporters of his claims, however, argue that this model is an obvious "plant" to discredit him.

Unidentified Submerged Objects and Underwater Bases

Unidentified Submerged Objects and Underwater Bases

3.13 Red Sphere in the Red Sea

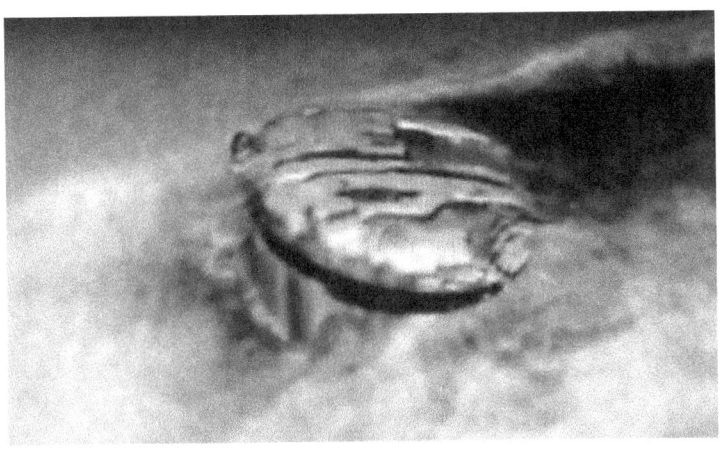

In August 1965, the crew of the Soviet steamship Raduga, while navigating in the Red Sea, observed a remarkable spectacle.

At around two miles away from their vessel, a fiery sphere dashed out from under the water and hovered over the surface of the sea illuminating the ocean in every direction. The sphere was estimated to be sixty meters in diameter, and it hovered above the sea at an altitude of 150 meters.

Unidentified Submerged Objects and Underwater Bases

Unidentified Submerged Objects and Underwater Bases

3.14 The Baltic Sea Anomaly

The Baltic Sea Anomaly is one mystery which has long bewitched UFO enthusiasts and tells of an alien spacecraft at the bottom of the sea

A gigantic pillar of water arose as the sphere emerged from the sea and then collapsed some moments later. This observation was mentioned in a number of Russian publications and was taken very seriously by the Soviet authorities. It is yet another typical example of a USO report.

Unidentified Submerged Objects and Underwater Bases

3.15 Silver Beehive

In the 1970s, reports issued by Admiral V.A. Domislovsky, chief of the Soviet Pacific Fleet's Intelligence Department, described an unknown, gigantic cylindrical object sighted by Soviet Navy in "faraway" regions of Pacific Ocean. The object was 800-900 meters long.

When it hovered over the ocean, smaller objects exited from one of its ends - like bees from a beehive - and descended into the waters.

Sometime later they re-entered the gigantic UFO. After the smaller objects thus "loaded" inside, the UFO flew away and disappeared over the horizon.

Unidentified Submerged Objects and Underwater Bases

Unidentified Submerged Objects and Underwater Bases

3.16 The Bermuda Triangle

The so-called Bermuda Triangle has been the sight of many strange events

According to a report on MosNews.com former Rear Admiral and nuclear submarine commander Yury Beketov saw USO activity inside the famous the Bermuda Triangle.

He said: "We repeatedly observed that the instruments detected the movements of material objects at unimaginable speed, around 230 knots (250 mph).
It's hard to reach that speed on the surface – only in the air is it readily possible.

'The beings that created those material objects significantly exceed us in development."

Russian Naval intelligence expert and Captain 1st Rank Igor Barklay also noted that the unidentified objects were

Unidentified Submerged Objects and Underwater Bases

most often spotted in deep water near where military forces are concentrated – off the Bahamas, Bermuda, Puerto Rico, and the east coast of the United States.

Unidentified Submerged Objects and Underwater Bases

3.17 The Devil's Sea

In the UK the Loch Ness monster mystery has sometimes been linked to aliens

Captain 1st Rank (now retired) Yuri Vinogradov who had served in the Soviet Navy from 1975 to 2000 also had a story to tell.

Vinogradov was a top expert in his field, had been involved in a number of submarine search and recovery operations; a veteran of "high-risk" units, and a participant in four long-range missions.

He had been to the Devil's Sea; also know as the Dragon Triangle which is located between Japan, Guam, and northern Philippines.

Some call this area the "Pacific Bermuda Triangle".

Unidentified Submerged Objects and Underwater Bases

In the 1980s, Vinogradov had participated in the search and rescue operations involving the Soviet Pacific Fleet.

Twice he and other officers had observed, on the sonar screen, a USO that had moved at great speed, and had disappeared in to the depths never to be seen again.

Unidentified Submerged Objects and Underwater Bases

3.18 Pyramid USOs over a Destroyer

Astonishing new footage captures 'pyramid UFOs swarming US Navy destroyer'

The amazing video and collection of photos of a "spherical" craft swarming a US Navy destroyer were allegedly part of a classified briefing conducted to "de-stigmatise the UAP problem"

Incredible footage supposedly taken by a US Navy destroyer has been leaked appearing to show several pyramid-shaped **UFOs** swarming above the ship at night.

Filmmaker Jeremy Corbell – who has produced documentaries for the likes of Netflix – was anonymously sent the baffling video from July 2019 as well as detailed information on an alleged intelligence briefing conducted by the Pentagon into Unidentified Aerial Phenomena (UAP).

On his Instagram and **website**, he claims he has been able to verify their authenticity having interviewed "those who would have knowledge of the events" as well as working with renowned investigative journalist George Knapp.

The 18-second clip taken onboard the USS Russell starts by showing what appears to be three faint orbs hovering in the sky.

The camera then pans to show a fourth object before zooming in on it to reveal its apparent triangular shape.

It also seems to be flashing repeatedly and is surrounded by two other fainter shapes – although it is not clear if they are simply lens flares.

As well as the eye-opening video, Jeremy shared three photos taken by the USS Omaha of an unidentified "spherical" craft.

"It is noted that the 'spherical' craft was suspected to be a transmedium vehicle and was observed descending into the water without destruction," he wrote.

Unidentified Submerged Objects and Underwater Bases

"It is noted that the 'spherical' craft could not be found upon entry to the water - and that a submarine was used in the search."

The two incidents appear to tally up with a previous story run by **The Drive** of multiple destroyers being swarmed by "mysterious drones" off California over a number of nights.

The producer was also sent information on a third sighting on March 4, 2019 from an FA-18 pilot and a Weapons Systems Officer.

He claimed the WSO took "personal mobile phone images of UAPs" out of the cockpit of the fighter jet.

The three incidents were allegedly spoken about during a classified briefing conducted by the Office of Naval Intelligence (ONI) about the UFO/UAP presence on May 1, 2020, according to Jeremy.

"Those familiar with the briefing articulated to me that the goal was to de-stigmatise the UAP problem and to promote more intelligence collection regarding UAP incursions and encounters with active military deployments," he said.

Speaking on his website, Jeremy explained the lengthy process he went through to "verify" the materials he received, before adding: "George Knapp and I were able to verify the materials you are about to consider. I can confirm their authenticity - as well as - the narrative supplied to me when they were presented.

"These are authentic photos and video from actual military encounters with UFOs - generated to educate high-level intelligence officers within our military on the nature and presentation of the UAP / UFO phenomenon."

He concluded that he hoped the materials "will be representative of a unique moment in modern history; a possible turning point towards the rational and transparent approach of investigating and exploring the mystery of the UFO Phenomenon".

Unidentified Submerged Objects and Underwater Bases

4.0 Other Interesting Stories- The Swimmers

An artist's impression of The Swimmers

One of the most peculiar of all such underwater encounters is one known as 'The Swimmers'.

In the summer of 1982 Mark Shteynberg, along with Lt. Colonel Gennady Zverev Soviet Navy), conducted periodic training of the reconnaissance divers from the Turkestan and Central Asian military regions.

Some of these training exercises took place at the Issyk Kul Lake in Krygyzstan.

Quite unexpectedly, the officers were paid a visit by a very important official, Major-General V. Demyanenko, commander of the Military Diver Service of the Engineer Forces of the Ministry of Defence.

Unidentified Submerged Objects and Underwater Bases

He informed the local officers of an extraordinary event that had occurred during similar training exercises in the Trans-Baikal and West Siberian military regions.

There, during their military training dives in Lake Baikal, the frogmen had encountered mysterious underwater swimmers, very human-like, except that their size was much larger--almost three meters tall.

Despite icy-cold water temperatures and a depth of fifty meters, they were dressed only in tight-fitting silvery suits with neither scuba diving equipment, nor any other equipment - only sphere-like helmets concealing their heads.

The local military commander, who was quite alarmed by such encounters, decided on a plan to capture one of the creatures.

To complete the mission, a special group of seven divers, under the command of an officer, had been dispatched.

Apparently, as the Soviet Navy frogmen tried to cover the creature with a net, a powerful unknown force threw the entire group out of the deep waters and up to the surface.

Unidentified Submerged Objects and Underwater Bases

5.0 Summary

These reports of USOs and Alien underwater bases are really amazing. There are many fewer USO reports than UFO reports on land, but that is to be expected given how many fewer people there are in the Oceans and Lakes at any one time.

It is not surprising when you think about it since 71% of the Earth is covered by water.

Could these sightings be evidence of a powerful technological undersea civilization here on our Earth? Possibly, since there is so little we know about things in the Oceans and lakes of our world.

I'm assuming we are talking about aliens from other worlds since being under the water would be a great place to hide.

In any event, it is interesting to speculate. The more I learn and research, the more I realize just how many mysteries there are in our world.

Martin Ettington

April 2021

Unidentified Submerged Objects and Underwater Bases

Unidentified Submerged Objects and Underwater Bases

6.0 Bibliography

1. https://www.popularmechanics.com/military/weapons/a29417939/unidentified-submarine-objects/. *Unidentified Submerged Objects.* [Online]

2. https://www.aetv.com/shows/the-lowe-files/exclusives/lowe-cation-underwater-alien-base. *Underwater Base off Malibu.* [Online]

3. https://www.thedrive.com/the-war-zone/25784/what-u-s-submariners-actually-say-about-detection-of-so-called-unidentified-submerged-objects. *Submariners and USOs.* [Online]

4. https://siberiantimes.com/other/others/features/f0077-aliens-and-ufos-at-worlds-deepest-lake/. *USOs in Lake Baikal.* [Online]

5. https://www.the-sun.com/news/1601315/russian-submarines-battle-aliens-oceans-kremlin-documents-2/. *Russian battle with USO.* [Online]

6. https://www.the-sun.com/news/1601315/russian-submarines-battle-aliens-oceans-kremlin-documents-2/. *Soviet Battles with USOs.* [Online]

7. https://www.tampabay.com/news/science/Alien-spaceship-may-lay-off-Florida-coast-says-Discovery-Channel-treasure-hunter_170882935/. *Alien Base off Florida.* [Online]

8. https://sports.yahoo.com/news/marine-claims-ufos-flew-out-of-underwater-base-121954944.html. *UFO of Cuba.* [Online]

9. https://blurbsurfer.com/video/c76rugc2. *UFO Video over Puerto Rico.* [Online]

10. https://www.mirror.co.uk/news/weird-news/secret-army-ufos-plotting-war-8634790. *UFOs Secret Army.* [Online]

11. https://www.dailystar.co.uk/news/weird-news/astonishing-new-footage-captures-pyramid-23882803. *New Footage of Pyramid UFOs swarming a destroyer.* [Online]

12. https://www.nippon.com/en/japan-topics/g00879/. *The Utsuro Bune Story.* [Online]

Unidentified Submerged Objects and Underwater Bases